I clip and snip scraps.

1

Cat sits still and naps.

I craft a cap!

"Can you snip?" I ask Cat.

Cat sits so still and naps.

3

I snap the color kit lid.

I grip a tan color.

"Can you color?" I ask Cat.
Cat still naps.

I color the cap.

It is Cat's cap!

I skip to Cat and slip on
the cap.

"Can you nap in a cap?" I ask.

Cat's cap slips as Cat scats.

The End